SpringerBriefs in Optimization

SpringerBriefs in Optimization showcases algorithmic and theoretical techniques, case studies, and applications within the broad-based field of optimization. Manuscripts related to the ever-growing applications of optimization in applied mathematics, engineering, medicine, economics, and other applied sciences are encouraged.

For further volumes:
http://www.springer.com/series/8918

Duc A. Tran

Data Storage for Social Networks

A Socially Aware Approach

 Springer

Duc A. Tran
Department of Computer Science
University of Massachusetts
100 Morrissey Blvd
Boston, MA
USA

ISSN 2190-8354   ISSN 2191-575X (electronic)
ISBN 978-1-4614-4635-4   ISBN 978-1-4614-4636-1 (eBook)
DOI 10.1007/978-1-4614-4636-1
Springer New York Heidelberg Dordrecht London

Library of Congress Control Number: 2012943345

Mathematics Subject Classification (2010): 91D30, 00-02, 97A40

Printed on acid-free paper

Springer is part of Springer Science+Business Media (www.springer.com)

Preface

Online social networking has become one of the most important forms of today's communication. While socially interesting features can make a system attractive, its competitive edge will diminish if it is not able to keep pace with increasing user activities. Deploying more servers is an intuitive way to make the system scale, but for the best performance one needs to determine where best across the servers to put the data, whether replication is needed, and, if so, how.

The purpose of this brief book is to introduce a novel approach to social data storage, which optimizes the distribution and replication of the data based on not only how often they are accessed but also their social relationships. The book is structured as follows. Chapter 1 provides a review of selected key technologies that form the data storage infrastructure for many of today's online social networks. Chapter 2 introduces optimization problems for socially aware data partitioning and replication, two of the most important design issues of data storage. Chapters 3 and 4 present recent solutions to these problems, respectively. The book is concluded with an epilogue.

Although related literature is reviewed, this book is not intended to be a comprehensive reference, but more as a resource advocating an unexplored area of research and development. Advanced students, software engineers, and computer scientists who work in the areas of networking and distributed data systems are the best audience for the book.

The author would like to thank his students, Khanh Nguyen and Ting Zhang, for help with the review of the existing technologies in Chap. 1 (Khanh) and the numerical results discussed in Chap. 3 (Ting) and Chap. 4 (Khanh) of this book.

Boston, Massachusetts, USA   Duc A. Tran

Contents

Chapter 1
Introduction

Evidenced by the success of Facebook with more than 800 million users, Twitter with more than 200 million users, and other sites alike, online social networks (OSNs) have become ubiquitous, offering novel ways for people to access information and communicate with each other. Nielsen published stats in June 2010 [28] showing that three of the world's top ten popular brands online are social-media related and, for the first time ever, social network or blog sites are visited by 75% of global consumers who go online. Among mobile users, social networking would surpass voice as the most popular form of mobile communication by 2015, according to Airwide Solutions [2].

The increasing popularity of social networking is undeniable, and so scalability is an important issue for any OSN that wants to serve a large number of users. There are two main ways to address scalability: vertical scaling and horizontal scaling. While vertical scaling scales "up" the system by adding more hardware resources to the existing servers, horizontal scaling scales "out" the system instead, by adding commodity servers and partitioning the workload across these servers. Vertical scaling is simple to manage, but horizontal scaling is more cost-effective and avoids the single-point-of-failure and bottleneck problems. The latter has been a de facto standard when it comes to managing data at massive scale for many OSNs today.

In a distributed storage system, where the data is partitioned across a number of servers, the data can also be replicated to provide a high degree of availability in case of failures. While data partitioning and replication is a well-known problem in the literature of distributed database systems [1, 14, 20, 33, 35–37], OSNs represent a novel class of data systems. In OSNs, a data read for a user often requires fetching the data of its neighbors in the social graph (e.g., friends' status messages in Facebook or connections' updates in LinkedIn). We refer to this property as *social locality*. Recent works [9, 32] have shown that distributed queries of small data records reduce performance compared to local queries. Therefore, social locality should be taken into account when designing a distributed storage for OSNs.

D.A. Tran, *Data Storage for Social Networks: A Socially Aware Approach*, SpringerBriefs in Optimization, DOI 10.1007/978-1-4614-4636-1_1, © Duc A. Tran 2012

More specifically, data of the users who are socially connected should be stored on servers within short reach from each other. An ideal storage scheme should be *socially aware*.

However, the most prominent distributed storage scheme for OSNs, Cassandra [21], is not socially aware. Originally deployed for Facebook to enhance its Inbox Search feature and now an Apache project, Cassandra has been used by most popular OSNs including Facebook, Twitter, Digg, and Reddit. While there exist well-known distributed file and relational database systems such as Ficus [33], Coda [36], GFS [14], Farsite [1], and Bayou [37], these systems do not scale with high read/write rates which is the case for OSNs. Cassandra's purpose is to be able to run on top of an infrastructure of many commodity storage hosts, possibly spread across different data centers, with high write throughput without sacrificing read efficiency. Cassandra is a key-value store resembling a combination of a BigTable data model [8] running on an Amazon's Dynamo-like infrastructure [11]. The data partitioning scheme underlying both Cassandra and Dynamo is based on consistent hashing [17], using an order-preserving DHT. Data is hashed to random servers, thus breaking their social locality.

Aimed to improve system performance and scalability by enforcing social locality in the data storage, socially aware data partitioning and replication schemes have been proposed. SPAR [32] is one such scheme that preserves social locality *perfectly* by requiring every two neighbor users to have their data colocated on the same servers. Since this is impossible if each user has only one copy of its data, replicas are introduced and placed appropriately in the servers such that the number of replicas needed to ensure perfect social locality is minimum. Another scheme is SCHISM [9], which can partition and replicate user data of a social graph efficiently by taking into account transaction workload such as how often two users are involved in the same transaction. In this book, we introduce two socially aware techniques for data partitioning and replication, S-PUT and S-CLONE, which our research group has recently devised. S-PUT is specifically designed for data partitioning, whereas S-CLONE is for data replication that assumes an existing data partition across the servers; this underlying data partition can be any arbitrary partition, e.g., a result of running Cassandra or S-PUT. Unlike SPAR and SCHISM, S-CLONE attempts to maximize social locality under a fixed space budget for replication. S-PUT and S-CLONE can be deployed separately or work together in a distributed storage system.

While later chapters of the book are focused on the issue of social locality, formulation of socially aware data partitioning and replication as optimization problems, and discussion of S-PUT and S-CLONE, this chapter is devoted to a brief review of Dynamo, BigTable, and Cassandra, three key techniques that form the data storage infrastructure for most OSNs today. The review is a summary of the details about these techniques as described in their source publications.

1.1 Amazon's Dynamo

Dynamo [11] is one of Amazon.com's database systems [others include SimpleDB, Simple Storage Service (S3)], used to manage a large amount of structured data stored across many commodity servers. The main data of interest is that of user shopping carts. It is a key-value store, the simplest form of NoSQL databases where data is stored as pairs of (key, value). NoSQL [6] represents an emerging class of non-relational database management systems suitable for data-intensive applications running across a large number of servers with heavy read/write activities. Because NoSQL does not require fixed table schemas and can scale horizontally without involving expensive join operations, it has been deployed in most popular OSNs.

In Dynamo, consistent hashing [17] is used to assign data to the storage servers. Data consistency is achieved using object versioning [22] with a quorum-like protocol governing updates. Error detection is implemented in a gossip-based manner. A system built on top of Dynamo is easy to administer and scale. With explicit membership control, adding or removing nodes has minimal effect on the overall system because nodes can bootstrap themselves into the system. Dynamo requires 99.5% requests to be returned within 300 ms. The core techniques implemented in Dynamo include partitioning, replication, versioning, read/write, and dynamics handling.

1.1.1 Partitioning

Dynamo employs *consistent hashing* to divide data among the servers. The idea is to organize the servers as nodes in a circular space, called a ring, where each server is given a random value in this space representing its position on the ring. Each data item, identified by a key, is assigned to a coordinator node by hashing this key to yield its position on the ring, and then walking the ring clockwise to find the first node right after the item's position. Thus, each node becomes responsible for the data items hashed to the region in the ring between it and its predecessor node. A read query or a write query of a user is always sent to its coordinator node.

Consistent hashing could lead to skewed partitions because it does not take into account the heterogeneity of nodes. Nodes with more resources may be underload while nodes with less resources may be overload. To cope with load imbalance, Dynamo introduces the concept of *virtual nodes*. Instead of having only one position on the ring, a node is assigned multiple positions called tokens. The number of tokens given to a node is determined based on factors such as its disk space and network bandwidth. Consequently, each node is responsible for multiple ranges of keys, each range starting at one of its tokens and ends at the predecessor corresponding to this token.

Fig. 1.1 Dynamo ring: key
K is primarily stored at B, its
coordinator node, and
replicated at C and D, the
two successor nodes of B;
here, we assume three copies
per data item

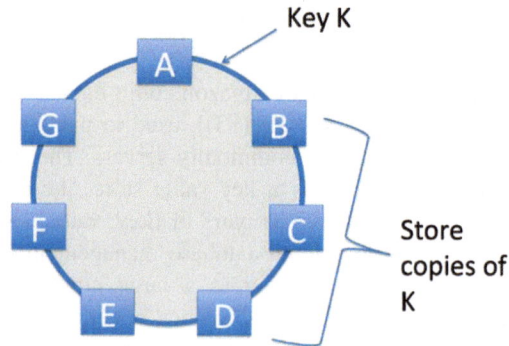

1.1.2 Replication

Dynamo allows data replication for better data availability,. Given a data item, in addition to storing its key locally, the coordinator node also replicates it on the $N-1$ successor nodes clockwise on the ring, where N is a pre-specified number representing the number of copies for the data item. For example, in Fig. 1.1, assuming $N=3$, node B, the coordinator node of key K, will replicate key K on node C and D, the two successor nodes of B on the ring. Consequently, each node is responsible for its range and the ranges of the $N-1$ predecessor nodes.

The list of nodes responsible for storing a key is called the key's *preference list*. To account for failure during write operations, this preference list may contain more than N nodes. Also, due to the presence of virtual nodes, it is possible that a key is replicated on two virtual nodes on the ring which happen to be the same physical node. Therefore, the preference list is constructed such that it includes only physical nodes but not virtual nodes. The system is designed so that any node can determine the preference list of a key based on exchanging membership and status information between nodes.

1.1.3 Versioning

Because data is replicated, data versioning is needed to achieve consistency. Dynamo provides *eventual consistency* in which write operations are applied to replicas asynchronously. As one of the consistency models in parallel computing, eventual consistency means that given a sufficiently long period of time over which no changes are sent, all updates are expected to propagate eventually throughout the system and so eventually all the replicas will be consistent.

Dynamo allows multiple versions of a data item to be available in the system at the same time. For example, on amazon.com, an "add to cart" operation can never be rejected nor forgotten. If for any reason, the most recent state of the cart is unavailable, the system must present the user with an older version of the cart. If the user makes a change to this old version, this change is still meaningful and must not be discarded when the most recent state of the cart becomes available again. The resolution for having multiple versions at the same time can be done either manually or automatically. A new version of an object should replace an older version and the system fixes itself, but when semantic contexts are involved the user must manually resolve the conflict. Dynamo uses vector clock [22] to maintain the evolution of the versions of an object.

1.1.4   Read and Write

Since there may be multiple versions for the same data, read and write in Dynamo follows a quorum-like protocol to ensure correctness in the data returned. This protocol has two configurable parameters: (1) R: the minimum number of nodes that must participate in a read operation, and (2) W: the minimum number of nodes that must participate in a write operation. The coordinator, which usually is the top node of the reference list, should handle read and write operations for a key. If a request is received at a random node that is not on the preference list, the request will be routed towards the list's top. If the request is for a write, the coordinator routes the request to the top N nodes in the preference list (N is the number of copies for a data item). If at least $W - 1$ nodes respond successfully, the write is considered successful. Similarly for a read request, the coordinator asks N nodes for the key and waits for at least $R - 1$ nodes to return data. If there is inconsistency in object versions, reconciliation is done manually or automatically depending on the nature of the conflict.

1.1.5   Handling Dynamics

Adding a node is relatively straightforward. A new node is assigned a set of tokens on the ring and, consequently, a set of key ranges it will be responsible for. In the case of a node's outage, this could be a permanent departure or happen due to a temporary failure or maintenance task. An outage is considered temporary unless a command is explicitly executed to inform the system of its permanent removal. Dynamo employs a gossip-based protocol to propagate changes in node membership, in which each node exchanges membership information with random neighbors once every second and thus can reconcile any inconsistency automatically. This periodic exchange of information allows nodes to be aware of the ranges managed by other nodes on the ring, so that a request arriving at any node can be forwarded directly to the right set of nodes responsible for this key.

To ensure read and write operations to never fail due to temporary node or network outages, Dynamo handles failures with *hinted handoff*. For example, when data needs to be replicated on a node that is currently not alive, the data will be written to a random node to be stored. This random node receives a hint indicating that the replica needs to be transferred back to the intended node when it is back online.

Hinted handoff is sufficient when the failure is transient. There are cases where a hinted replica becomes unavailable before it can be returned to the intended node. To address this problem, Dynamo uses an *anti-entropy* protocol to keep the replicas synchronized. In this protocol, each node maintains a separate Merkle tree [25] for each key range it is responsible for and allows other nodes to verify if a key within a key range is up to date. The advantage of Merkel tree are twofold. First, each branch of the tree can be checked independently without having to download the entire tree. Second, it is easy to compare if two trees are the same by comparing the hash values at their roots. Therefore, Merkle tree can minimize the amount of data needed to transfer when checking for replica inconsistencies. In Dynamo, if two nodes share a key range, they exchange the roots of the Merkel trees. Having identical root values implies no inconsistency; all the replicas are up-to-date. In the case of different root values, the nodes keep exchanging hash values of the children recursively until the leave nodes of the trees are reached. At that point we can identify the replica that is stale and perform synchronization accordingly.

1.2 Google's BigTable

BigTable [7] is a schema-free distributed high performance database system built on Google's technologies such as Google File System, Chubby Lock Service [4], and Sorted String Table (SSTable). BigTable does not provide support for a full relational model; instead, it allows dynamic control over data layout. BigTable is richer than key-value and can be used to manage sparse and semi-structured data.

1.2.1 Table Format

A BigTable table is essentially a sparse persistent multidimensional sorted map. Each cell value in the map is stored as an array of bytes and indexed by a triple *(row-key, column-key, time-stamp)*. Every value is associated with a timestamp t_i. Timestamp is a 64 bit integer used to recognize different revisions of a cell value. BigTable orders these values in decreasing order of timestamp so that the newest version is always read first.

Row key is a string of up to 64 KB in size. Rows are lexically ordered and grouped into *tablets*. When the data is distributed across the servers, it is the tablets that are actually distributed. Data access, therefore, can take advantage of the data locality in each tablet to reduce the number of servers needed to contact during read/write operations.

There is no limit on the number of columns in each table. Each row can have an arbitrary number of columns. While rows are grouped into tablets, columns are grouped into *column families* each consisting of related columns. Column family is an essential concept of BigTable's data model; hence, BigTable often referred to as a column-oriented datastore. A column family is predefined within a table. However, the number of columns in a column family can be arbitrary and may change going from row to row.

1.2.2 Distributed Storage

BigTable stores data and log files on top of GFS. The data is stored in Google's SSTable file format. A SSTable is an ordered string-to-string map whose key and value are arbitrary. Given an SSTable, applications can look up a value via its key or iterate over all key-value pairs in the table. BigTable uses Chubby [4], a distributed persistent lock service, to handle operations such that routing requests, locating tablets, and storing metadata like column families for each table.

There are three important entities in BigTable's immplementation: the master server, tablet servers, and the client library: There is a master server for each BigTable instance, which is responsible for assigning tablets to their corresponding servers (called tablet servers), adding and removing tablet servers, and balancing workloads among tablet servers. Although the master server is centralized, its load is low because most clients do not contact it for tablet lookup. Instead, the tablet location information is handled by Chubby. When a master starts up, it creates a special file into a Chubby namespace and acquires an exclusive lock. If the master cannot hold the lock or its session expires, it takes itself down. Thus, the availability of a BigTable instance relies on the reliablity of the connection between the master server and the Chubby service.

Tablet servers can be added and removed during run time. A tablet maintains an unassigned status until the master finds a server capable to serve it. Similar to the master server, when a tablet server starts, it places a file in the Chubby namespace and acquires an exclusive lock on it. The tablet server can serve its assigned tablets as long as it can hold the lock with Chubby. The master server periodically asks each tablet server if it still holds the lock. If the table server does not respond, the master checks with Chubby if the tablet server still holds the lock on its file. If this fails, the master deletes the file and marks all the tablets on the failed server unassigned.

The client library provides an interface for the application to interact with BigTable. It is responsible for looking up tablet servers, directing requests to them, and returning results to the application.

1.3 Apache Cassandra

Cassandra [21], initially developed at Facebook and later becoming an Apache open source project, is a data store solution resembling a combination of Dynamo and BigTable. Cassandra can be described as one data store that runs a BigTable data model on a Dynamo-like server infrastructure. Cassandra is arguably the most popular choice for implementing a large-scale distributed storage system today. It has been used by Digg, Facebook, Twitter, Reddit, Rackspace, Cloudkick, and Cisco, to name a few. Below, we discuss some key features of Cassandra.

1.3.1 Data Model

A "database"—the core concept of relational databases—is called a *keyspace* in Cassandra. Analogous to a database containing a set of relations or tables, a keyspace of Cassandra contains a set of *column families*. Like relational tables, column families must be defined when the Cassandra keyspace is created and they cannot be modified thereafter. Adding or removing column families requires reboot of the keyspace. However, the fundamental difference between a relational table and a column family is that while the former is composed of rows with the same columns, different rows of a column family do not have to share the same columns. A row can have any number of columns and these columns can vary from row to row. For example, Fig. 1.2 shows a Cassandra column family with two rows, one with three columns and one with two columns.

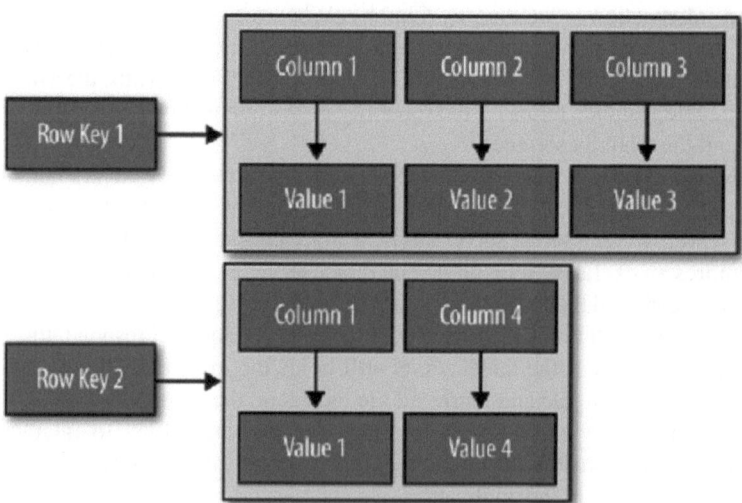

Fig. 1.2 Cassandra's table with column families (source: [16])

A column value is a pair of name-value associated with a timestamp; e.g., using the Java Script Object Notation (JSON) representation, we can have a column named "Name,"

```
{
    name: "Name",
    value: "Duc A. Tran",
    timestamp: 123456789
}
```

and a column named "Address,"

```
{
    name: "Address",
    value: "100 Morrissey Blvd",
    timestamp: 125555555
}
```

A column is the smallest increment of data and is identified by its name. A column family is simply a set of rows, each composed of an arbitrary number of columns. For example, we can have a column family named "User Profile" below:

```
UserProfile = {
    name: "User Profile",
    User1: {
       Name: {
        name: "Name",
        value: "Duc A. Tran",
        timestamp: 123456789
       },
       Genre: {
        name: "Genre",
        value: "M",
        timestamp: 123456789
       },
       Address: {
        name: "Address",
        value: "100 Morrissey Blvd",
        timestamp: 125555555
       },
       Phone: {
        name: "Phone",
        value: "617-287-6452",
        timestamp: 123456789
       }
    },
    User2: {
       Name: {
```

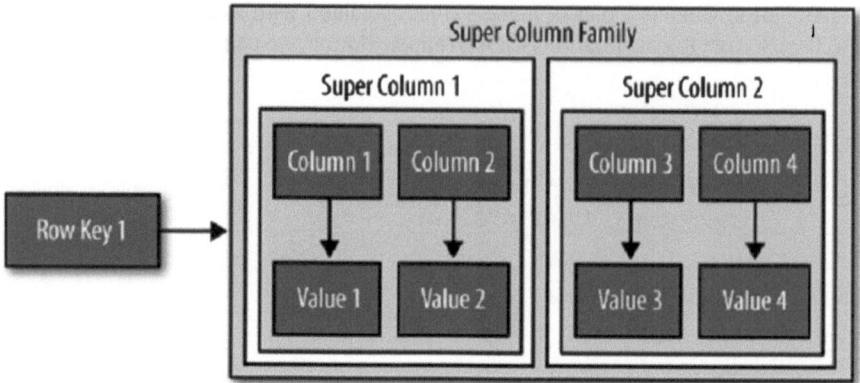

Fig. 1.3 Cassandra's super column (source: [16])

```
        name:  "Name",
        value:  "Jane Doe",
        timestamp:  125555555
        },
      Address:  {
        name:  "Address",
        value:  "25 Main Street",
        timestamp:  125555555
        },
      Genre:  {
        name:  "Genre",
        value:  "F",
        timestamp:  125555555
        }
     }
  }
}
```

In this example, the column family has two rows, identified by row keys "User1" and "User2," respectively.

Each column family is stored in a separate file and therefore the columns of a family should be those that are frequently accessed together. Row key in Cassandra serves a role similar to a primary key in relational databases to join tables together. It is important to note that row keys are entered by the client, the purpose being that in order to store column families across a number of hosting servers in a distributed manner, row key is used to determine which server to store the data.

Cassandra also provides another layer of data abstraction: super columns and super column families (Fig. 1.3). A super column is a name-value pair like a standard column but the value of a super column is a map of columns, not an atomic value. Unlike a row in a regular column family which is basically a sorted map of {*column*

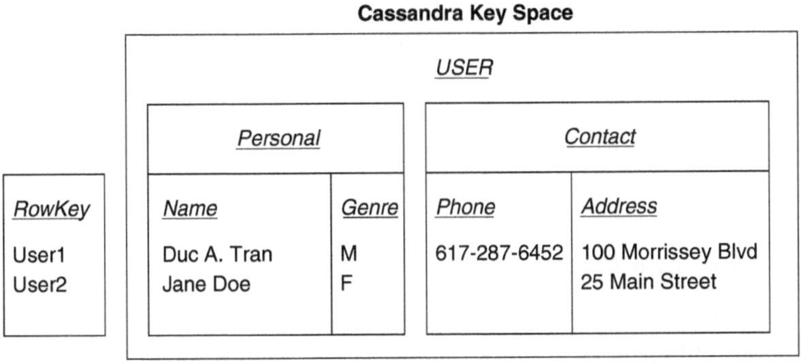

Cassandra Key Space

USER				

	Personal		*Contact*	
RowKey	*Name*	*Genre*	*Phone*	*Address*
User1	Duc A. Tran	M	617-287-6452	100 Morrissey Blvd
User2	Jane Doe	F		25 Main Street

Fig. 1.4 Cassandra table with a super column family

names → *column values*}, a row in a super column family is a sorted map of {*super column names* → *maps of column names to column values*}. Figure 1.4 shows a super column family named *"USER,"* which consists of two super columns, *"Personal"* and *"Contact."*

1.3.2 Partitioning

Similar to Dynamo, Cassandra uses consistent hashing based on user ID to partition the data across the storage nodes. Each node is assigned a unique position, or a *token*, on the ring, and responsible for a range of row keys starting from the predecessor node's token to this node's token. Cassandra differs from Dynamo in dealing with load balancing due to the heterogeneity of the nodes. As discussed in Sect. 1.1.1, Dynamo uses the concept of virtual nodes to improve load balancing. Cassandra does not implement this approach. Instead, it analyzes the load information at each physical node and performs redistribution of load whenever the system detects an imbalance. This way, Cassandra wants to make the design and implementation tractable and provide deterministic decisions regarding load balancing.

1.3.3 Replication

Cassandra allows the application to choose its replication policy on top of the data partition. One policy provided by Cassandra, namely "Rack Unaware," is to replicate each data item on the successor nodes of its coordinator node on the ring. This approach is similar to that of Dynamo. Other replication policies are "Rack Aware" and "Data Center Aware" which take into account the load balancing across the servers within a data center, as well as across multiple data centers. For example, in the "Rack Aware" policy, the first replica is placed at the first node on the ring that belongs in another data center, and the remaining replicas are placed at the first nodes on the ring in the same rack as the first. Succeeding nodes on the ring should alternate data centers to avoid hot spots.

1.3.4   Handling Dynamics

Dynamics handling in Cassandra is similar to that in Dynamo with slight modifications. Cassandra employs Scuttlebutt [34], an anti-entropy, gossiped-based protocol for membership control. A node A considers a node B failed if A cannot reach B (even if B is still responsive to some other node C). Also, instead of having only two status values, "up" or "down," Cassandra calculates a value that represents a suspicion level for each node. The status of a node depends on a configurable threshold.

1.3.5   Querying

Cassandra provides an API of methods to access data, which is built entirely on top of Thrift, a software framework for scalable cross-language services development (http://thrift.apache.org). Some important functions are below:

- $insert()$: insert a column given name and value
- $get()$: retrieve by column name for a key
- $multiget()$: retrieve by column name for a set of keys
- $get\_slice()$: retrieve columns or super columns by column name or a range of names for a key
- $get\_range\_slice()$: a subset of columns for a range of keys
- $multiget\_slice()$: a subset of columns for a set of keys; similar to $get\_range\_slice()$ but allow non-contiguous keys

For a social network, a read request usually involves reading data for multiple users (the requesting user and its friends), requiring a $multiget()$ call. If data storage does not preserve social locality, this call may result in having to contact multiple servers, which is less efficient than if only one single server is contacted to return all the relevant data. We discuss this issue in detail in the next chapter.

Chapter 2
Social Locality in Data Storage

The *locality* property in data storage can be interpreted in different ways. In Cassandra a column family is a group of columns that are frequently accessed together, e.g., name, address, phone number, and email address information. These columns therefore have the same row key resulting in their being stored on the same machine. Data locality of this kind is content-based. By social locality, we are refering to the data that are accessed by users that share some social relationship. Therefore, although these data may be content-wise unrelated, they are frequently queried together in an online social network and therefore should be stored in close proximity on disk. Another way to look at locality is in terms of geography. It may be desirable to store in the same server the data for those users that reside in the same geographic region (e.g., think Akamai).

Although content- and geography-based locality have been taken into account in the literature of data storage, social locality is an emerging concept. Social locality is not achieved by Cassandra which uses random-based consistent hashing to assign data to servers. To illustrate the benefit of social locality, consider a 10-node social graph in Fig. 2.1a stored across three servers, A, B, and C, in a distributed manner. Suppose that using Cassandra partitioning we have the partition shown in Fig. 2.1b. We allow one replica for each user in addition to the primary copy. Figure 2.1c shows the result of using a random replication algorithm, such as the Rack Unaware strategy of Cassandra, on top of this partition; this algorithm basically randomly places the replicas among the three servers. In contrast, Fig. 2.1d shows the result of running an (imaginary) replication algorithm that preserves social locality, trying to place data of neighbor nodes on the same server as much as possible. Table 2.1 summarizes the cost to read the data for each of the ten users, showing a noticeable improvement of socially aware replication over random replication (24% better). This example, albeit its simplicity, supports the importance of social locality in distributed data storage.

D.A. Tran, *Data Storage for Social Networks: A Socially Aware Approach*, SpringerBriefs in Optimization, DOI 10.1007/978-1-4614-4636-1_2, © Duc A. Tran 2012

Fig. 2.1 Random vs. socially aware (three servers, one primary and one replica per user node): replicas are shown in *dashed circles* and primaries in *solid circles*. (**a**) Social graph (**b**) Primary partition (**c**) Random replication (**d**) Socially-aware replication

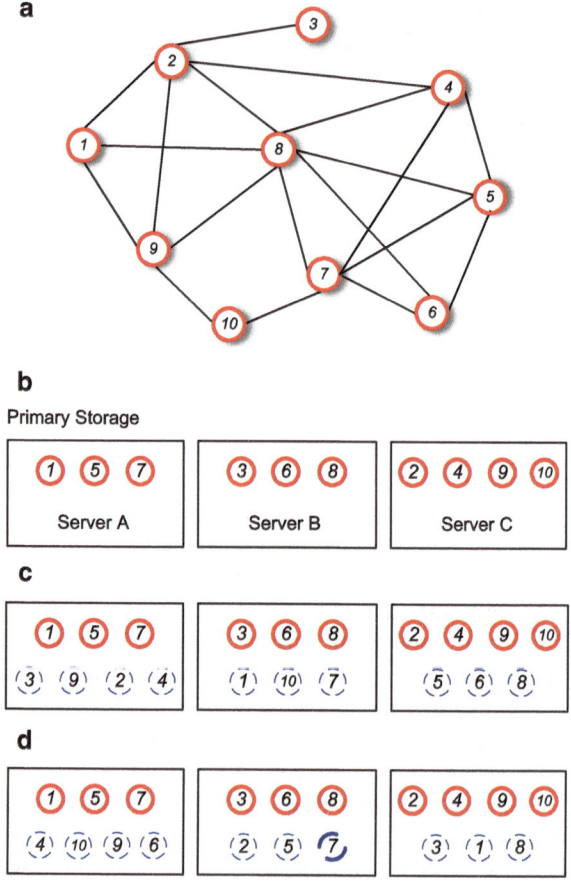

Table 2.1 Socially-aware vs. random replication: read cost of each user

Method/User	1	2	3	4	5	6	7	8	9	10	Total
Random	2	3	1	2	3	2	4	4	2	2	25
Socially-aware	3	1	1	3	2	1	2	3	1	2	19

2.1   Perfect vs. Imperfect Social Locality

Ideally, the preservation of social locality in a storage system should be *perfect*, i.e., every user must have its data located on the same server with all of its neighbor users. Consequently, any read request by a user asking for its data and that of its neighbors can be fulfilled efficiently by visiting only one server. Perfect social locality, however, is not always possible if the number of replicas for each user is limited by some upper bound. Therefore, there are different design problems depending on the specific requirements of the storage system under consideration. For a system that requires perfect social locality, we may want to minimize the number of replicas needed. For a system that has a fixed disk budget for the replicas, and hence, *imperfect* social locality is the only choice, we may want to maximize the (partial) preservation of social locality. In either case, server workloads (read and write) and load balancing are additional factors that need to be addressed.

In the following sections, we discuss a select set of optimization problems for socially aware storage. We start in the next section with some assumptions and notations.

2.2 Assumptions and Notations

Consider a storage system with M servers to store data for a social graph of N users. We assume that each user's data is of small size and as such it is desirable to minimize the number of servers required to access a given number of data records. Illustrated in Fig. 2.2, we assume a three-tier system architecture,

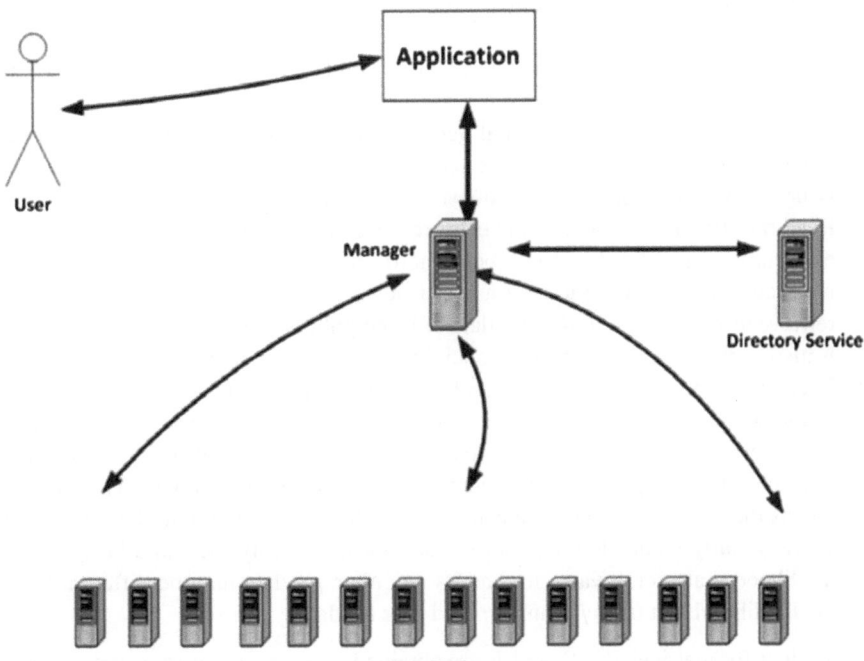

Fig. 2.2 Three-tier storage system architecture

User-Manager-Server, in which the users do not communicate with the servers directly. Instead, the *Manager*, providing API and directory services, serves as the interface between the users (frond end) and the servers (back end). The API is used for the users to issue read and write requests. Assisted by the directory service, each request for a user is always directed to its primary server who is responsible for fulfilling the request on the user's behalf. The directory service can be implemented either as a global map or as a DHT. The connectivity information of the social graph is assumed to be available at the Manager.

The system is characterized by the following parameters:

- Partition assignment P: a $N \times M$ binary matrix representing the primary assignment of user data across the servers. In this matrix, each entry p_{is} has value 1 if and only if user i is assigned to server s. The mapping from a server to a user is surjective and, therefore,

$$\sum_{s=1}^{M} p_{is} = 1 \forall i \in [1, N] \tag{2.1}$$

- Replication assignment X: a $N \times M$ binary matrix representing the replica assignment of user data across the servers. In this matrix, each entry x_{is} has value 1 if and only if user i is replicated at server s. Since a replica cannot reside on the same server with its primary copy, we have

$$x_{is} + p_{is} \leq 1 \forall (i, s) \in [1, N] \times [1, M] \tag{2.2}$$

- Write rate W: an N-dimensional vector representing user write request rates. Each element w_i is a real positive number quantifying the rate at which user i issues a write request. A write request for a user is always sent to its server.
- Read rate R: an N-dimensional vector representing user read request rates. Each element r_i is a real positive number quantifying the rate at which user i issues a read request. A read request for a user is always first to its server, requesting to retrieve its data and *probably* the data of its neighbors. Whether a neighbor's data is also retrieved is determined by social bond strength, which is defined below.
- Social relationship E: an $N \times N$ real matrix representing the social relationships in the social graph. In this matrix, each entry e_{ij} is a value in the range $[0, 1]$ quantifying the social bond between user i and user j. A stronger social bond indicates stronger probability (tendency) to read each other's data. The value 1 means the strongest and 0 means no relationship. It is noted that although i and j are socially connected, e_{ij} and e_{ji} are not necessarily the same because the likelihood that user i wants to read its neighbor j's data may be different from the likelihood that user j wants to read user i's data.

The values for matrices W, R, and E are obtained based on monitoring and analysis of actual workload.

2.3 Optimization Objectives

Minimizing the server load and maximizing load balancing are among the most important objectives of any distributed storage system. The server load is categorized into three types: read load, write load, and storage load. We formulate these objectives below.

2.3.1 Read Load

To understand the server read load , consider a read request for user i. This request needs to be directed to its primary server, say server s, which will provide the data for i. Suppose that user i is also interested in the data of user j, one of its neighbors. To retrieve this data, there are two cases:

- User j's data (primary or replica) is located on server s: The data of j can be provided by server s, requiring no additional server request.
- User j's data (primary or replica) is not located on server s: The data of j needs to be retrieved from the primary server of j, requiring one additional read request sent to this server.

The amount of data returned to user i is the same in both cases, but the number of read requests that need to be processed at the server side is different, and, especially, worse if i and j do not colocate. In OSNs where data under request are of small size, having more read requests incurs more traffic and CPU processing at the server side. Therefore, an important objective is to minimize the server load due to read requests, which we refer to as *read load*, given the social relationships between the users and the rate at which they initiate read requests.

Given a server s, its read load is computed as

$$l_s^{\text{read}} = \sum_{i=1}^{N} r_i \left(p_{is} + (1 - p_{is}) \sum_{j=1}^{N} p_{js} e_{ij} \sum_{t=1}^{M} p_{it}(1 - x_{jt}) \right). \quad (2.3)$$

This load consists of read requests that are initiated by (1) users i primarily assigned to s and (2) users i not primarily assigned to s, that have neighbors j primarily assigned to s but these neighbors are not colocated with i. The number of read requests belonging to the latter group depends on the social strength matrix E which determines whether a neighbor's data needs also to be retrieved.

The total read load of all the servers is

$$L^{\text{read}} = \sum_{s=1}^{M} l_s^{\text{read}} = \sum_{s=1}^{M} \sum_{i=1}^{N} r_i \left(p_{is} + (1 - p_{is}) \sum_{j=1}^{N} p_{js} e_{ij} \sum_{t=1}^{M} p_{it}(1 - x_{jt}) \right)$$

$$(2.4)$$

which can be re-written as

$$L^{\text{read}} = \underbrace{\sum_{i=1}^{N} r_i \left(1 + \sum_{j=1}^{N} e_{ij} \sum_{s=1}^{M}(1-p_{is})p_{js}\right)}_{L_0^{\text{read}}} - \underbrace{\sum_{i=1}^{N} r_i \sum_{s=1}^{M}(1-p_{is}) \sum_{j=1}^{N} p_{js}e_{ij} \sum_{t=1}^{M} p_{it}x_{jt}}_{\Delta}$$

(2.5)

(derived using the equality $\sum_{s=1}^{M} p_{is} = 1 \forall i \in [1, N]$). Note that L_0^{read} represents the total read load in the case without replication (i.e., $X = 0$) and Δ represents the read load reduction offered by the replication scheme X. Letting s_i denote the primary server of user i, i.e., $p_{is_i} = 1$ and $p_{is} = 0 \; \forall \; s \neq s_i$, we can simplify Δ as follows

$$\Delta = \sum_{i=1}^{N} r_i \sum_{s=1}^{M}(1 - p_{is}) \sum_{j=1}^{N} p_{js}e_{ij} \sum_{t=1}^{M} p_{it}x_{jt}$$

$$= \sum_{i=1}^{N} r_i \sum_{s \neq s_i}^{M} \sum_{j=1}^{N} p_{js}e_{ij} x_{js_i}$$

$$= \sum_{i=1}^{N} r_i \sum_{j=1}^{N} x_{js_i} e_{ij} \sum_{s \neq s_i}^{M} p_{js}$$

$$= \sum_{i=1}^{N} r_i \sum_{j=1}^{N} x_{js_i} e_{ij} (1 - p_{js_i})$$

$$= \sum_{i=1}^{N} r_i \sum_{j=1}^{N} x_{js_i} e_{ij} - \sum_{i=1}^{N} r_i \sum_{j=1}^{N} x_{js_i} e_{ij} p_{js_i}$$

$$= \sum_{i=1}^{N} r_i \sum_{j=1}^{N} x_{js_i} e_{ij} \text{ (because } x_{js_i} \text{ and } p_{js_i} \text{ cannot both be 1)}$$

$$= \sum_{i=1}^{N} r_i \sum_{j=1}^{N} \sum_{s=1}^{M} p_{is}x_{js}e_{ij}$$

$$= \sum_{j=1}^{N} \sum_{s=1}^{M} x_{js} \sum_{i=1}^{N} r_i p_{is}e_{ij}$$

$$= \sum_{i=1}^{N} \sum_{s=1}^{M} x_{is} \sum_{j=1}^{N} r_j p_{js}e_{ji}.$$

The last derivation is the result of switching index i with index j. Hence, we have

$$L^{\text{read}} = \sum_{i=1}^{N} r_i \left(1 + \sum_{j=1}^{N} e_{ij} \sum_{s=1}^{M} (1 - p_{is}) p_{js} \right) - \sum_{i=1}^{N} \sum_{s=1}^{M} x_{is} \sum_{j=1}^{N} r_j p_{js} e_{ji}. \quad (2.6)$$

As visible in the above formulas, to minimize the total read load and/or balance the read load across the servers, we have to take into account how the data is partitioned (P) and replicated (X) and the social relationships among the users (E).

2.3.2 Write Load

A write request initiated by a user requires update on its primary copy as well as all the replicas. We refer to the number of write requests a server has to process as its *write load*, which depends on the number of users for whom the server stores data (primary or replica) and the rate at which they initiate write requests.

Given a server s, its write load is computed as

$$l_s^{\text{write}} = \sum_{i=1}^{N} w_i \left(p_{is} + x_{is} \right). \quad (2.7)$$

The total write load of all the servers is

$$L^{\text{write}} = \sum_{s=1}^{M} l_s^{\text{write}} = \sum_{s=1}^{M} \sum_{i=1}^{N} w_i \left(p_{is} + x_{is} \right) \quad (2.8)$$

$$= \sum_{i=1}^{N} w_i \left(\sum_{s=1}^{M} p_{is} + \sum_{s=1}^{M} x_{is} \right) \quad (2.9)$$

$$= \sum_{i=1}^{N} w_i \left(1 + \sum_{s=1}^{M} x_{is} \right). \quad (2.10)$$

While the write load of a server depends on how the data are partitioned and replicated, the total write load depends only on the replication scheme (X) regardless of the partitioning scheme (P). It is noted that the social relationships (E) do not have any impact on write load.

2.3.3 Storage Load

We compute the storage load of a server as the number of users whose data, primary or replica, is stored at this server. Given a server s, its storage load is computed as

$$l_s^{\text{store}} = \sum_{i=1}^{N} (p_{is} + x_{is}).$$ (2.11)

The total storage load of all the servers is

$$L^{\text{store}} = \sum_{s=1}^{M} l_s^{\text{store}} = \sum_{s=1}^{M} \sum_{i=1}^{N} (p_{is} + x_{is})$$ (2.12)

$$= \sum_{i=1}^{N} \left(\sum_{s=1}^{M} p_{is} + \sum_{s=1}^{M} x_{is} \right)$$ (2.13)

$$= N + \sum_{i=1}^{N} \sum_{s=1}^{M} x_{is}$$ (2.14)

which depends only on the total number of replicas for all the users in the system regardless of which servers they are located. Similar to write load, storage load is not affected by the social relationships.

2.3.4 Load Balancing

To represent the degree of load balancing across the servers, a variety of measures of statistical dispersion can be used, such as coefficient of variation, standard deviation, mean difference, and Gini coefficient. For example, recently, coefficient of variation has been used in the work of [27, 32, 38] and Gini coefficient in [15, 29–31].

Coefficient of variation (CV) is defined as the ratio of the standard deviation σ to the mean μ

$$\text{CV} = \frac{\sigma}{\mu}$$

which can only apply to measurements of positive values. CV is independent of the unit in which the measurement is taken and useful in cases where the standard deviation of data must always be understood in the context of the mean of the data. Distributions with CV < 1 are considered low variance, while those with CV > 1 high variance.

Whereas CV can get very large with small mean, Gini coefficient can be used to compare load balancing of different distributions independent of their size, scale, and absolute values. This measure naturally captures the fairness of the load distribution, with a value of 0 expressing total equality and a value of 1 maximal inequality. Supposing that y_s represents the load of server s, be it read load, write load, or storage load, the Gini coefficient representing the inequality of server load is

$$G = \frac{2}{(M-1) \sum_{i=1}^{M} y_s} \sum_{s=1}^{M} \bar{s} y_s - \frac{M+1}{M-1}$$ (2.15)

Table 2.2 Server loads (individual and total)

Description	Notation	Formula
Individual read load	l_s^{read}	$\sum_{i=1}^{N} r_i (p_{is} + (1 - p_{is}) \sum_{j=1}^{N} p_{js} e_{ij} \sum_{t=1}^{M} p_{it}(1 - x_{jt}))$
Total read load	L^{read}	$\sum_{i=1}^{N} r_i (1 + \sum_{j=1}^{N} e_{ij} \sum_{s=1}^{M} (1 - p_{is}) p_{js}) - \sum_{i=1}^{N} \sum_{s=1}^{M} x_{is} \sum_{j=1}^{N} r_j p_{js} e_{ji}$
Individual write load	l_s^{write}	$\sum_{i=1}^{N} w_i (p_{is} + x_{is})$
Total write load	L^{write}	$\sum_{i=1}^{N} w_i (1 + \sum_{s=1}^{M} x_{is})$
Individual Storage load	l_s^{store}	$\sum_{i=1}^{N} (p_{is} + x_{is})$
Total storage load	L^{store}	$N + \sum_{i=1}^{N} \sum_{s=1}^{M} x_{is}$

where $\bar{s}$ denotes the rank of server s based on the ascending order of load y_s. The range of G is $0 \leq G \leq 1$. To balance the server load, we need to minimize this Gini coefficient.

In the next section discussing formulating the data partitioning and replication as multi-objective optimization problems, we use Gini coefficient to represent load balancing. This is for the convenience of presentation; depending on how solutions are approached, other measures can also be used.

2.4 Multi-Objective Optimization

Ideally, we want to simultaneously minimize the total read load, total write load, and total storage load of all the servers while balancing individual loads across the servers. These objectives, however, are conflicting with each other. For example, we could place all the data on the same server to minimize the read load, but this would incur severe imbalance of the storage load. Also, the write load and storage load cannot be balanced at the same time, nor can they be minimized, because of the write rate vector W. Thus trade-offs are inevitable and we should be specific about which objectives are given more priority before we design the system.

In what follows, we formulate the optimization problem for data partitioning and data replication. This formulation is generic in the sense that when applied to a certain system the objectives and constraints can be modified accordingly to fit its specifications. For ease of lookup, we summarize the general formulas for server loads in Table 2.2.

2.4.1 The Partitioning Problem

In this problem, we assume no replication and are required to partition the N users across the M servers such that the given optimization objectives are satisfied. Without replicas, i.e., $x_{is} = 0$, the server loads are computed as follows:

- Load of a server

$$l_s^{\text{read}} = \sum_{i=1}^{N} r_i \left(p_{is} + (1 - p_{is}) \sum_{j=1}^{N} p_{js} e_{ij} \right) \tag{2.16}$$

$$l_s^{\text{write}} = \sum_{i=1}^{N} w_i \, p_{is} \tag{2.17}$$

$$l_s^{\text{store}} = \sum_{i=1}^{N} p_{is} \tag{2.18}$$

- Total load

$$L^{\text{read}} = \sum_{i=1}^{N} r_i \left(1 + \sum_{j=1}^{N} e_{ij} \sum_{s=1}^{M} (1 - p_{is}) p_{js} \right) \tag{2.19}$$

$$L^{\text{write}} = \sum_{i=1}^{N} w_i \tag{2.20}$$

$$L^{\text{store}} - N \tag{2.21}$$

Since the total write load and total storage load are constant regardless of the partitioning scheme, in addition to load balancing, we need only to minimize the total read load. The total read load is minimized if all the data are assigned to the same server, say server s, resulting in a total of $\sum_{i=1}^{N} r_i$. This case, however, incurs the worst load imbalance because every server except s has zero load and server s is the only server having to process all read and write requests. Therefore, in addition to minimizing the total read load, it is desirable to balance the server loads. Using Gini coefficient to quantify the degree of load balancing, the optimization can be modeled as the following multi-objective optimization problem:

Problem 2.1 (Partitioning). Find the best binary matrix P

$$\underset{P}{\text{minimize}} \quad \left[L^{\text{read}}, G^{\text{read}}, G^{\text{write}}, G^{\text{store}} \right]^T$$

$$\text{subject to} \quad \sum_{s=1}^{M} p_{is} = 1 \text{ for } 1 \leq i \leq N$$

where G^{read}, G^{write}, and G^{store} are the Gini coefficient of read load, write load, and storage load, respectively.

2.4.2 The Replication Problem

In this problem, we are given a partition which already assigns users to their corresponding servers (i.e., matrix P is known) and required to find the best replication assignment (X). As aforementioned in Sect. 2.1, we may have different objectives depending whether the system can afford perfect social locality or imperfect social locality.

2.4.2.1 Replication with Perfect Social Locality

With perfect social locality, every user must have its neighbor users primarily stored or replicated on its primary server. Read requests, therefore, do not have to go cross-server. Given a server s, its read load is incurred only due to read requests initiated by its primarily assigned users,

$$l_s^{\text{read}} = \sum_{i=1}^{N} r_i \, p_{is}.$$

Hence,

$$L^{\text{read}} = \sum_{s=1}^{M} \sum_{i=1}^{N} r_i \, p_{is}$$

$$= \sum_{i=1}^{N} r_i \sum_{s=1}^{M} p_{is}$$

$$= \sum_{i=1}^{N} r_i.$$

Individual read loads and the total read load are independent of the replication scheme. It turns out that there is only one replication scheme that satisfies the requirement for perfect social locality. This requirement can be represented by the following constraint

$$e_{ij} \, p_{is}(1 - p_{js})(1 - x_{js}) = 0 \text{ for } 1 \le i, j \le N, 1 \le s \le M$$

i.e., given a user i who is primarily located at server s (i.e., $p_{is} = 1$), for each neighbor j (i.e., $e_{ij} > 0$), we must have either j is primarily assigned to s (i.e., $p_{js} = 1$) or a replica of j is at s (i.e., $x_{js} = 1$).

Consequently, given a server s and a user j, there are two cases:

1. If j is primarily assigned to s (i.e., $p_{js} = 1$): we do not need to put a replica of j on s and therefore $x_{js} = 0$.

2. If j is not primarily assigned to s (i.e., $p_{js} = 0$): if there exists a neighbor i primarily assigned to s (i.e., $e_{ij} > 0$, $p_{is} = 1$), then $x_{js} = 1$; otherwise, $x_{js} = 0$.

In any case, we can formulate x_{js} as

$$x_{js} = (1 - p_{js}) \left(1 - \prod_{i=1}^{N}(1 - \lceil p_{is}e_{ij} \rceil) \right)$$

and so there is no optimization involved.

2.4.2.2 Replication with Imperfect Social Locality

Imperfect social locality is the case for systems with limited storage budget for replication. Also, although users are not equally active in the network, they should deserve the same degree of data availability so that everyone has an equal chance to successfully access data under any failure condition. Therefore, a useful constraint is to the same number, K, of replicas for each user, in addition to their primary copy; i.e.,

$$\sum_{s=1}^{M} x_{is} = K \text{ for } 1 \le i \le N. \tag{2.22}$$

With this constraint, the total write load and storage load can be simplified as

$$L^{\text{write}} = \sum_{i=1}^{N} w_i \left(1 + \sum_{s=1}^{M} x_{is} \right) = (K + 1) \sum_{i=1}^{N} w_i \tag{2.23}$$

$$L^{\text{store}} = N + \sum_{i=1}^{N} \sum_{s=1}^{M} x_{is} = N(K + 1) \tag{2.24}$$

Since these loads are fixed regardless of the replication scheme, we focus on minimizing the total read load and balancing read load, write load, and storage load. The problem, therefore, is modeled as follows:

Problem 2.2 (Replication). Find the best binary matrix X

$$\underset{X}{\text{minimize}} \quad \left[L^{\text{read}}, G^{\text{read}}, G^{\text{write}}, G^{\text{store}} \right]^{T}$$

subject to   1) $x_{is} \le 1 - p_{is}$ for $1 \le i \le N, 1 \le s \le M$

2) $\sum_{s=1}^{M} x_{is} = K$ for $1 \le i \le N$

2.4.3   Other Problems

There are other problems related to partitioning and replication of social data that have been addressed but not discussed in this chapter. Addressed in [32] is a problem aimed to find both the partitioning matrix P and replication matrix X such that the number of replicas is minimized under the requirement that perfect locality be achieved. The load balancing is taken into account, however, in a secondary role. This problem does not consider the heterogeneity of user read/write rates and social strengths.

Patterns of user activities are an important design factor for social data storage, as stated in the problem addressed in [5]. User rates and social strengths are computed based on a time window of past social activities. The main focus of this problem is data partitioning whose goal is to minimize and balance read load. Perfect locality is not required.

In another problem [9], transaction workload is taken as input to find a partitioning and replication placement with balanced load and minimal cross-server communication. The number of replicas for each data node is determined based on the number of transactions accessing the node. Hence, the number of replicas may change from one node to another and minimizing the total amount of replicas is not a goal.

The partitioning problem and replication problem we have formulated in the previous sections are different from these related problems. Our partitioning problem is similar to the problem of [5] but with different objective functions. Our replication problem has not been addressed in the literature.

Chapter 3
S-PUT

S-PUT is a socially aware partitioning framework that we propose as a solution to the partitioning problem (Problem 2.1) discussed in the last chapter. Ideally, we want to minimize the total read load of all the servers while simultaneously balancing the server loads, including read load, write load, and storage load. These objectives, however, are conflicting with each other. For example, we could place all the data on the same server to minimize the read load, but this would result in severe imbalance of the storage load. Also, the write load and storage load cannot be balanced at the same time because of the write rate vector W. Thus trade-offs are inevitable and we should be specific about which objectives are given more priority before the actual design.

S-PUT is specifically designed for systems that prefer low read load and balanced write load. This solution can easily be adapted to working with systems that want low read load and balanced storage load because the formulas for individual read load $l^{\text{store}}(s)$ and total read load L^{store} are just special cases of the formulas for individual write load $l^{\text{write}}(s)$ and total write load L^{write}, respectively. Indeed, by setting $W = \{1, 1, \ldots, 1\}$, $l^{\text{store}}(s)$ and L^{store} will exactly be $l^{\text{write}}(s)$ and L^{write}, respectively.

Equation (2.19) provides the formula for the total read load,

$$L^{\text{read}} = \sum_{i=1}^{N} r_i \left(1 + \sum_{j=1}^{N} e_{ij} \sum_{s=1}^{M} (1 - p_{is}) p_{js}\right).$$

To balance the write load, S-PUT aimes to minimize the Gini coefficient,

$$G^{\text{write}} = \frac{2}{(M-1) \sum_{i=1}^{N} w_i} \sum_{s=1}^{M} s \sum_{i=1}^{N} w_i p_{is} - \frac{M+1}{M-1}$$

(here, the servers have been ranked in increasing order of write load, $\sum_{i=1}^{N} w_i p_{is}$).

D.A. Tran, *Data Storage for Social Networks: A Socially Aware Approach*, SpringerBriefs in Optimization, DOI 10.1007/978-1-4614-4636-1_3, © Duc A. Tran 2012

Problem 2.1 is then re-expressed as follows.

Problem 3.1 (Partitioning with Minimum Read Load and Balanced Write Load). Find the best binary matrix P

$$\underset{P}{\text{minimize}} \quad \left[L^{\text{read}}, G^{\text{write}} \right]^T$$

$$\text{subject to} \quad 1) \sum_{s=1}^{M} p_{is} = 1 \text{ for } 1 \leq i \leq N$$

$$2) \sum_{i=1}^{N} w_i p_{is} \leq \sum_{i=1}^{N} w_i p_{it} \text{ for } 1 \leq s < t \leq M$$

3.1 Approach

Problem 3.1 belongs to the class of nontrivial multi-objective optimization problems for which we cannot identify a perfect solution optimizing all the objectives. To problems of this kind, an alternative goal is to look for a set of *Pareto-optimal* solutions. A Pareto-optimal solution is one that is not "dominated" by any other solution; solution A is dominated by solution B if A is worse than B in every objective and *strictly* worse in at least one objective.

The research area of multi-objective optimization originated in the late 1980s and several solution approaches have since been proposed [13]. An intuitive approach is to combine the conflicting objectives into a single weighted aggregate objective function which is easier to optimize. The optimality of this approach, however, is dependent on the selection of the weight coefficients in the aggregate function. How to choose the best values for these coefficients is a nontrivial task. Alternatively, an effective approach, which is based on evolutionary algorithms (EA), has often been used [41]. The EA-based approach after many generations of crossover and mutation can eventually result in a good set of Pareto-optimal solutions. The drawback of EA, however, is due to its slow convergence to a stable solution state, especially for large-size problems, which unfortunately is the case for our optimization problems with thousands of unknown variables.

S-PUT relies on the eventual guarantee of EA in finding a set of good partitioning solutions. However, to avoid the long convergence time, instead of starting with a random solution candidate set, the EA used in S-PUT starts with a population of candidate solutions resulted from applying a classic graph partitioning algorithm. It is noted that one could apply a graph partitioning algorithm such as [3, 18, 26] to obtain a data partition with improved social locality. This approach is employed in [5, 9]. However, a graph partitioning algorithm alone may not produce a solution that is optimized for all the objectives and constraints of our optimization problem. With S-PUT we show that such an algorithm can provide a good set of initial partition assignments that can later be significantly improved upon by EA.

3.2 Algorithm

The S-PUT algorithm consists of two phases. In the *Initial Partitioning* phase, a graph partitioning algorithm is used to obtain the initial population for the EA process. In the *Final Partitioning* phase, the EA process takes place to result in the final set of optimized partitioning assignments.

3.2.1 *Initial Partitioning*

It is observed that if we denote $f_{ij} = r_i e_{ij} + r_j e_{ji}$, the total read load can be expressed as

$$L^{\text{read}} = \sum_{i=1}^{N} r_i + \frac{1}{2} \times \sum_{i=1}^{N} \sum_{j=1}^{N} f_{ij} \sum_{s=1}^{M} (1 - p_{is}) p_{js}.$$

Thus, to minimize L^{read}, we need to minimize

$$\sum_{i=1}^{N} \sum_{j=1}^{N} f_{ij} \sum_{s=1}^{M} (1 - p_{is}) p_{js}.$$

This quantity is the sum of f_{ij} of pairs of socially-connected users, i and j, who are assigned to different servers. Consider an undirected weighted graph G formed by the vertices and links of the original social graph, where each vertex i of G is associated with a weight w_i and each link (i, j) of G with a weight f_{ij}. Our optimization problem is equivalent to finding an optimal partitioning of graph G into M components such that (1) the edge cut, i.e., sum of the weights of inter-component links, is minimum (to minimize L^{read}), and (2) the total vertex weight of each component is balanced (to satisfy the balancing constraint). This is one of constrained weighted graph partitioning problems known to be NP-hard [3, 18, 23, 26], but there are well-approximated heuristic algorithms. Among them is METIS [18], arguably the best approximate algorithm for partitioning a large graph into equally weighted components with minimum edge cut. METIS can partition a 1-million-node graph in 256 parts in just a few seconds on today's PCs.

In S-PUT, we first obtain a set of partition assignments, each being a result of applying METIS on graph G using a different "seed." Seed is a random factor used in METIS; a different assignment can be obtained by using a different value for the seed. This set of assignments is used to populate the first generation of EA. The cardinality of this set is equal to the "population size," an input parameter for the EA process, explained in what follows.

3.2.2 Final Partitioning

EA is an iterative process of generations, starting with an initial population in the first generation and iteratively improving the population from one generation to the next. The evolution of the population is driven by three main mechanisms: recombination, mutation, and selection. Recombination and mutation create the necessary diversity, thus facilitating novelty in the population. After recombination and mutation, a selection step takes places to select the best-quality individuals for the next-generation's population. A fitness function is used to determine the quality of each individual.

Among many evolutionary algorithms, the Non-dominated Sorting Genetic Algorithm-II (NSGA-II) [10] and Strength Pareto Evolutionary Algorithm 2 (SPEA2) [40] are de facto for solving multi-objective optimization problems. Although either can work in S-PUT for its evolution process, here we describe this process using SPEA2, which we have evaluated with some encouraging preliminary results.

Using SPEA2 as EA for the partitioning problem, a population is represented by a set of individuals, each being a base-M string of length N, $s_1 s_2 \ldots s_N$, representing a possible partition assignment: user i is assigned to server s_i. For example, for a network of 1,000 nodes to be partitioned across 16 servers, an individual is an array of 1,000 integers, each having a value between 0 and 15. For the recombination mechanism, which creates two new individuals, called offsprings, replacing two parent individuals selected randomly from the current population, we use the Two-Point Recombination Mode (the other modes provided by SPEA2 are One-Point Recombination and Uniform Recombination). Suppose that the two parents are $s_1 s_2 \ldots s_N$ and $s_1' s_2' \ldots s_N'$. First, two random positions in the string are chosen, i and j ($1 < i < j < N$). Then, the offsprings are $s_1 \ldots s_{i-1} s_i' \ldots s_j' s_{j+1} \ldots s_N$ and $s_1' \ldots s_{i-1}' s_i \ldots s_j s_{j+1}' \ldots s_N'$. For the mutation mechanism, a new offspring is created replacing a parent individual by assigning a random value to each of a number of random positions in the parent array. Hence, an offspring of $s_1 s_2 \ldots s_N$ can be $s_1 .. s_{i-1} t s_{i+1} \ldots s_N$ if one bit needs to be mutated, where $i \in [1, N]$ and $t \in [0, M-1]$ are both chosen at random. When applied to a population of size $size$, the number of parent couples in the recombination step and the number of single parents in the mutation step are size $\times$ $p_{\text{recombination}}$ and size $\times$ p_{mutation}, respectively, where $p_{\text{recombination}}$ and p_{mutation} are probabilities given as input parameters to the EA. For mutation, the number of bits (i.e., genes in biological terminology) that need mutating is $N \times p_{\text{gene\_mutation}}$ where $p_{\text{gene\_mutation}}$ is another probability serving as input for the EA.

SPEA2 maintains two types of population, $\mathscr{P}_h$ (called the "regular population," size $|\mathscr{P}|$) and $\mathscr{A}_h$ (called the "archive," size $|\mathscr{A}|$), for each generation h, which are updated, basically, as follows.

1. First generation ($h = 0$): $\mathscr{P}_0$ is the initial population of $|\mathscr{P}|$ individuals (result of applying METIS with $|\mathscr{P}|$ different seeds) and $\mathscr{A}_0$ is set to empty.
2. Generation ($h + 1$):

    a. $\mathscr{A}_{h+1}$ = set of non-dominated individuals of $\mathscr{P}_h \cup \mathscr{A}_h$ (truncated to $|\mathscr{A}|$ individual if the population size is larger than $|\mathscr{A}|$, or padded with lowest-fitness individuals among the dominated if the population size is less than $|\mathscr{A}|$)
    b. $\mathscr{P}_{h+1}$ = set of $|\mathscr{P}|$ individuals after application of recombination and mutation on $\mathscr{A}_{h+1}$

When the maximum number of generations, h^*, is reached, the final partition assignments will be the non-dominated individuals in the final archive $\mathscr{A}_{h^*}$. An individual p dominates an individual q, denoted by $\mathsf{p} \succ \mathsf{q}$ if the corresponding partition assignment of p is no worse than that of q in terms of both L^{read} and G^{write}, with at least one objective *strictly* better. The fitness of an individual p in generation h is defined as

$$\text{fitness}(\mathsf{p}) = \sum_{\mathsf{q} \in \mathscr{P}_h \cup \mathscr{A}_h : \mathsf{q} \succ \mathsf{p}} \text{strength}(\mathsf{q}) + \frac{1}{\sigma_{\mathsf{p},k} + 2}$$

where *strength*(q) is the number of individuals dominated by p in the set $\mathscr{P}_h \cup \mathscr{A}_h$ and $\sigma_{\mathsf{p},k}$ denotes the distance in the objective space from individual p to its k-nearest individual. As common setting, k is set to $\sqrt{|\mathscr{P}| + |\mathscr{A}|}$.

3.3 Numerical Results

We present here the results of a simulation study to evaluate S-PUT with SPEA2, using the Facebook dataset made available by Max-Planck Software Institute for Software Systems [39]. This dataset contains a Facebook network of $N = 63,392$ users in New Orleans region, with 816,886 links resulting an average degree of 25.7. We assume that the read and write rates of a user are proportional to its social degree, thus these rates are given values in the range $(0, 1)$ proportional to the degree. The social strength between two neighboring nodes is uniformly generated, also in the range $(0, 1)$.

The number of servers is set to $M = 16$. The parameters for SPEA2 are: regular population size $|\mathscr{P}| \in \{100, 500\}$, archive size $|\mathscr{A}| = |\mathscr{P}|$, recombination probability $p_{\text{recombination}} = 0.8$, mutation probability $p_{\text{mutation}} = 0.5$, gene mutation probability $p_{\text{gene\_mutation}} = 0.001$, and number of generations $h^* = 100$.

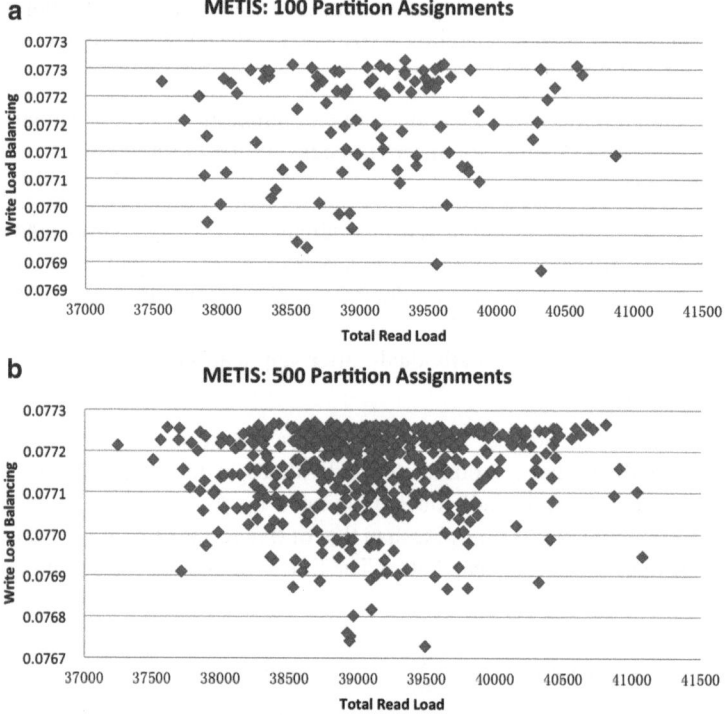

Fig. 3.1 Initial partition solutions as a result of METIS. (**a**) 100 individuals (**b**) 500 individuals

Figure 3.1 shows the partition solutions resulted from METIS as described in Sect. 3.2.2. When using these solutions for the initial population in the EA process of SPUT, the results obtained in after 100 generations are plotted in Fig. 3.2, which shows a significant improvement. For the case with 100 individuals Fig. 3.2a, we obtain twelves non-dominated solutions, all no worse than METIS. Ten out of these solutions are substantially better than METIS solutions in terms of both the toal read load (L^{read}) and write load balancing (G^{write}). Similarly for the case with 500 individuals Fig. 3.2b, we obtain eight non-dominated solutions, five of which are substantially better than METIS in both measures.

These results shows that METIS cannot provide good solutions to our partitioning problem and substantiates the effective use of EA in order to reach a set of solutions of significantly better quality. In terms of run time, it took about 6 h to finish the S-PUT program for the case of 500 individuals.

3.4 Notes

S-PUT is a socially aware partition scheme aimed at minimizing the total read load and balancing the write load. S-PUT can be modified easily to work with systems that require low read load and balanced storage load instead of balanced write

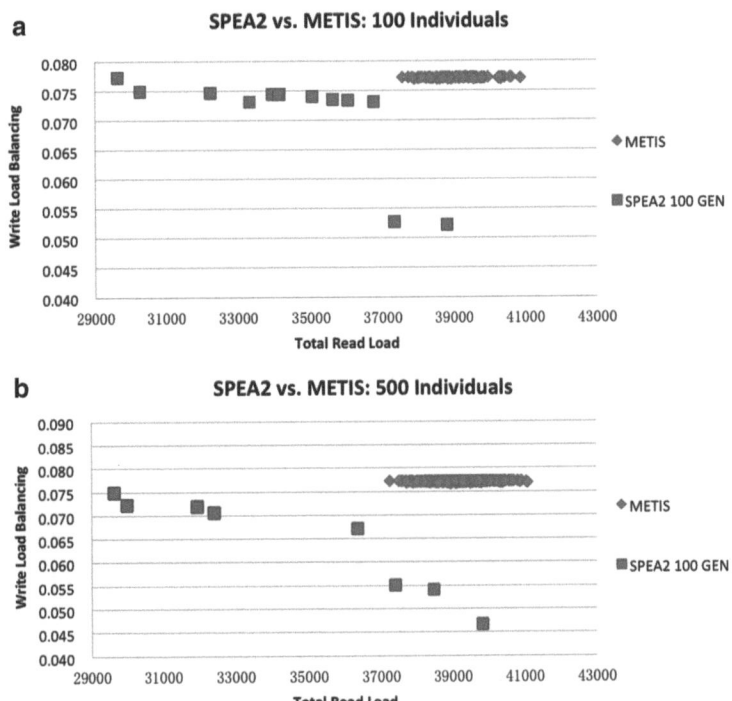

Fig. 3.2 Final solutions after SPEA2 is applied to an initial population of solutions provided by METIS. (**a**) 100 individuals (**b**) 500 individuals

load. The optimization of S-PUT takes into account the user read/write activity and social relationship. Although S-PUT relies on an evolutionary algorithm, its convergence time is shortened significantly by starting with an initial population of decent solutions provided by METIS, an efficient partitioning algorithm for large graphs.

Random partitioning based on consistent hashing such as that in the case of Cassandra has the advantage of being efficient because there is no need to maintain a global map telling which server is assigned to each user. Given the user id, the request can be routed to the right server without knowing in advance the identity of this server. In S-PUT, the identity of this server needs to be known in advance, which is obtained from a directory service. This directory service can be made efficient by being implemented as a DHT. S-PUT is most suitable for deployment in a cluster of servers where a given server can be reached in one hop.

Chapter 4
S-CLONE

S-CLONE is a solution we propose to the replication problem, Problem 2.2, discussed in Chap. 2. S-CLONE can run on top of any given data partition and is applicable to OSNs that want a fixed budget for the disk space and update cost required for replication. It provides the same degree of data availability for every user so that everyone has an equal chance to successfully access data under any failure condition.

As discussed in earlier chapters, although it is ideal, it is unrealistic to minimize the total read load of all the servers while simultaneously balancing the server loads, including read load, write load, and storage load. These objectives are conflicting with each other. S-CLONE is especially aimed at minimizing the read load as the primary objective and balancing the write load as the secondary objective. Focusing on only these objectives, Problem 2.2 is re-expressed as follows:

Problem 4.1 (Replication with Minimum Read Load and Balanced Write Load). Find the best binary matrix X

$$\underset{X}{\text{minimize}} \quad \left[L^{\text{read}}, G^{\text{write}}\right]^T$$

subject to 1) $x_{is} \leq 1 - p_{is}$ for $1 \leq i \leq N, 1 \leq s \leq M$

2) $\sum_{s=1}^{M} x_{is} = K$ for $1 \leq i \leq N$

3) $\sum_{i=1}^{N} w_i(p_{is} + x_{is}) \leq \sum_{i=1}^{N} w_i(p_{it} + x_{it})$ for $1 \leq s < t \leq M$

D.A. Tran, *Data Storage for Social Networks: A Socially Aware Approach*, SpringerBriefs in Optimization, DOI 10.1007/978-1-4614-4636-1_4, © Duc A. Tran 2012

where the formulas for L^{read} and G^{write} are shown again below for ease of reference:

$$L^{\text{read}} = \sum_{i=1}^{N} r_i \left(1 + \sum_{j=1}^{N} e_{ij} \sum_{s=1}^{M} (1 - p_{is}) p_{js} \right) - \sum_{i=1}^{N} \sum_{s=1}^{M} x_{is} \sum_{j=1}^{N} r_j p_{js} e_{ji}$$

and

$$G^{\text{write}} = \frac{2}{(M-1) \sum_{i=1}^{N} w_i (1+K)} \sum_{s=1}^{M} s \sum_{i=1}^{N} w_i (p_{is} + x_{is}) - \frac{M+1}{M-1}$$

(here, the servers have been ranked in increasing order of write load, $\sum_{i=1}^{N} w_i (p_{is} + x_{is})$).

4.1 Approach

The evolutionary algorithm (EA) approach has been shown in Chap. 3 as an effective way to obtain good partitioning assignments. Although EA could theoretically be applicable to the replication problem, its effectiveness would not be guaranteed due to the size of the solution space. While there are N^M partitioning assignments possible for the partitioning problem, there can be $N^{C(M,K)}$ possible replication assignments for the replication problem, a significant increase ($C(.)$ represents the number of K-combinations). Furthermore, the efficiency of S-PUT relies on the result of METIS, a good graph-partitioning algorithm. This is not applicable to the replication problem which is given an existing data partition to start with.

Instead of relying on a combination of EA and METIS as done in S-SPUT, S-CLONE adopts a greedy algorithm. It is observed that minimizing L^{read} is equivalent to

$$\text{maximize} \sum_{i=1}^{N} \sum_{s=1}^{M} x_{is} \sum_{j=1}^{N} r_j p_{js} e_{ji}$$

and so if we need to place a replica copy for a user i somewhere, the most desirable location should be the primary server of most neighbors of i, taking into their social strengths and read rates; this way, most neighbors will benefit from this replica when they issue a read query, thus helping to reduce the total read load. The detailed algorithm is described in the next section.

4.2 Algorithm

S-CLONE consists of two phases: *replicate* and *adjust*.

1. *Replicate* phase: The procedure starts in this phase. It works in a greedy manner, sequentially considering a node at a time and finding the best way to replicate its data. Suppose that the nodes are processed in the order $\{1, 2, \ldots, N\}$. For each node i (initially node 1) that has not been processed (i.e., all the nodes $1, 2, \ldots, i - 1$ have been processed), the K replicas are assigned to their corresponding servers as follows. First, the location histogram below is computed for i

$$h_i(s) = \sum_{j=1}^{N} r_j p_{js} e_{ji} \tag{4.1}$$

where $h_i(s)$ is the number of i's neighbor nodes who are primarily stored at server s. Then, the top K servers (with highest h_i values) are chosen to each store a replica for user i. If there is a tie, the server with the least write load (so far) is preferred. A special case occurs where there are not enough K servers in the histogram; in this case, the remaining replicas for node i will be placed in the *Adjust* phase below.

2. *Adjust* phase: The goal of this phase is to find the servers for the remaining replicas that cannot be placed due to the special case aforementioned above. Because the read cost is the same no matter where to store these replicas, their locations are chosen to maximize load balancing. The best way to do this is to process each remaining replica one by one and place it on the server that currently has the least load.

After the K replicas of i have been placed, the algorithm moves on to process the next user, node $(i + 1)$. Note that in choosing the K servers to replicate the data for a user, we do not consider its primary server because it makes no sense to put a replica and its primary on the same server. Also, as only the location information of the primary copies is used to determine where to replicate a user, the order to process the users does not have impact on the final replica placement. This is a desirable property because there may be different structures to represent a social graph and S-CLONE can work with any structure unbiasedly.

The algorithm above is the basic version of S-CLONE, which treats load balancing as the secondary objective. For systems that require a better degree of load balancing, we can extend S-CLONE by enforcing the following constraint:

$$\forall s : l_s^{\text{write}} = \sum_{i=1}^{N} w_i (p_{is} + x_{is}) \leq (1 + \varepsilon)(L^{\text{write}}/M) = (1 + \varepsilon)(K + 1) \sum_{i=1}^{N} w_i / M$$

i.e., the write load of each server cannot exceed the average write load by a factor of ε, a small value representing the desired load balancing. Here, the formulas for

l_s^{write} and L^{write} are obtained from Eqs. (2.7) and (2.23), respectively. The *Replicate* phase is the same as in the basic version of S-CLONE, except that in searching for the best server to replicate each user, we do not consider those servers that violate the above new constraint.

4.3   Numerical Results

In an effort to substantiate the efficiency of S-CLONE, a preliminary evaluation of S-CLONE has been conducted in [38]. This evaluation is applied to the basic version of S-CLONE for the special case where all users in the social graph are equally active (by setting $W = R = 1$) and the strength of a social relationship is either 0 (no relationship) and 1 (socially connected). We discuss this evaluation here as a way to illustrate the importance of social locality in data replication and by taking advantage of how S-CLONE performs in comparison with a random-based technique like Cassandra. The dataset used for the evaluation was obtained from the Max-Planck Software Institute for Software Systems [39], containing a Facebook network of $N = 63,392$ users in New Orleans region, with 816,886 links resulting an average degree of 25.7. This is the same dataset used in the evaluation of S-PUT discussed in the previous chapter. In the evaluation, the number of storage servers M varies in the set $\lfloor 8, 16, 32 \rfloor$, and the number of replicas K in $[1, M - 1]$.

   S-CLONE is evaluated on top of two partitions of the social graph, which are resulted from applying random partitioning and METIS partitioning, respectively. Random partitioning resembles a DHT-based key-value scheme *à la* Cassandra to randomly assign each user to a server. METIS partitioning is a graph partitioning scheme [19] that aims to divide a graph into K equal-sized components such that the social links are maximally preserved in each component. This method is the same method used in S-PUT discussed in the previous chapter. The purpose of evaluating with METIS for partitioning is to see how the replication schemes compare in the case where the partition itself does already preserve some degree of social locality.

   Figures 4.1–4.3 plot the read cost per user which is the ratio of the total read load to the number of users, for cases $M = 8$, $M = 16$, and $M = 32$, respectively. First, we discuss the case where replication is not used, i.e., each user has only one server location, which is its primary location. Using random partitioning to assign user data to the servers the average read cost is about 24 server reads per user when the number of servers is eight (Fig. 4.1a) and this cost increases slightly as more servers are deployed, to a cost of about 27 when there are 32 servers (Fig. 4.3a). If we use METIS partitioning, the average read cost is about 8 server reads per user when the number of servers is eight (Fig. 4.1b), reaching a cost of 12 with 32 servers (Fig. 4.3b).

   In the case where replication is used, with K replicas per user, the read cost improves. Regardless of the partitioning method underneath, random replication improves linearly with the number of replicas, whereas S-CLONE offers a more

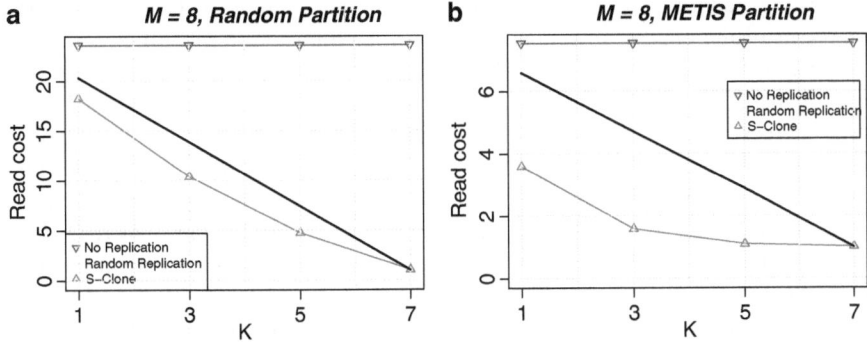

Fig. 4.1 Read cost of S-CLONE vs. random replication with $M = 8$ servers, (**a**) when applied on top of the random partition and (**b**) when applied on top of the METIS partition

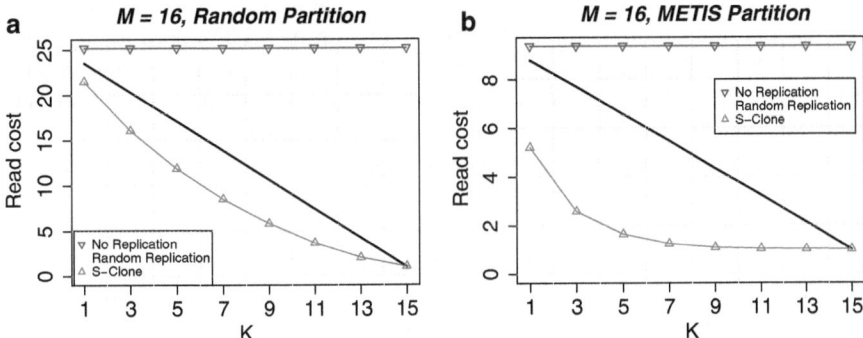

Fig. 4.2 Read cost of S-CLONE vs. random replication with $M = 16$ servers, (**a**) when applied on top of the random partition and (**b**) when applied on top of the METIS partition

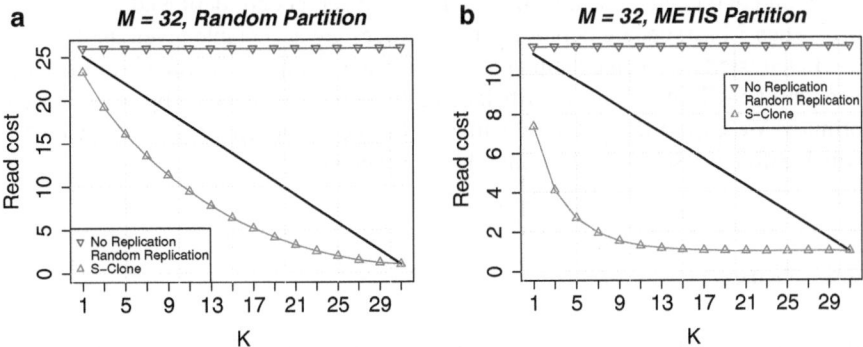

Fig. 4.3 Read cost of S-CLONE vs. random replication with $M = 32$ servers, (**a**) when applied on top of the random partition and (**b**) when applied on top of the METIS partition

interesting pattern. The improvement in S-CLONE is more significant in early increases of the number of replicas but not so significant after a sufficiently large number. For example, in the case $M = 32$ when S-CLONE is applied on top of the random partition (Fig. 4.3a), the read cost of S-CLONE drops quickly from 24 to 5 as K increases from 1 to 17, but afterwards the decrease is less significant. The drop is quicker if S-CLONE is applied on top of the METIS partition. This implies that although both the random partition and METIS partition offer comparable degrees of load balancing, to achieve the same improvement rate for the total read load, we need fewer replicas per user if METIS partitioning is used than if random partitioning is used. The reason, we conjecture, is because METIS does preserve social locality whereas random partitioning does not. Consequently, social locality should be considered highly in the storage design.

The superiority of S-CLONE to random replication is obvious, especially when more servers are deployed or when METIS is used for partitioning instead of random partitioning. For example, on top of random partitioning when $M = 32$, in order to achieve a read cost of 5, S-CLONE requires 17 replicas per user (i.e., $K = 11$) but random replication requires 26 replicas. On top of METIS partitioning when $M = 32$, S-CLONE requires just 3 replicas per user but random replication requires 19 replicas. It is thus important that we take social locality into account not only when we store the primary data, but also when we replicate it.

We also observe that, for each given M, there is a value for K that maximizes the efficiency gap between S-CLONE and random replication. For example, in the case $M = 32$ (Fig. 4.3a), this value is $K = 15$. The gap is narrower as K is approaching towards 1 or towards $M - 1$. This is understandable because in these two extreme cases there is no substantial difference in the replica placement using either partitioning scheme. It will be interesting though to derive a formula for the optimal value of K that will maximize the efficiency gap.

In terms of load balancing, Fig. 4.4 plots the Gini coefficient of S-CLONE for cases $M = 8$, $M = 16$, and $M = 32$ when it is applied on top of the random partition and the METIS partition. It is observed that S-CLONE balances the load better when more servers are deployed or when more replicas are allowed per user. The Gini coefficient is at most 0.35 when eight servers are deployed and at most 0.17 when 32 servers are deployed. These values are acceptable given the fact that S-CLONE starts with an existing partition and the results are obtained for the basic version of S-CLONE with load balancing being the secondary objective, not the primary. We expect better Gini coefficient for the enhanced version of S-CLONE which enforces a stricter constraint on load balancing

4.4 Notes

For OSNs that already employ an arbitrary data partition structure, whose data need to be replicated, we can increase the extent of social locality during the replication procedure. S-CLONE is a socially aware replication scheme which,

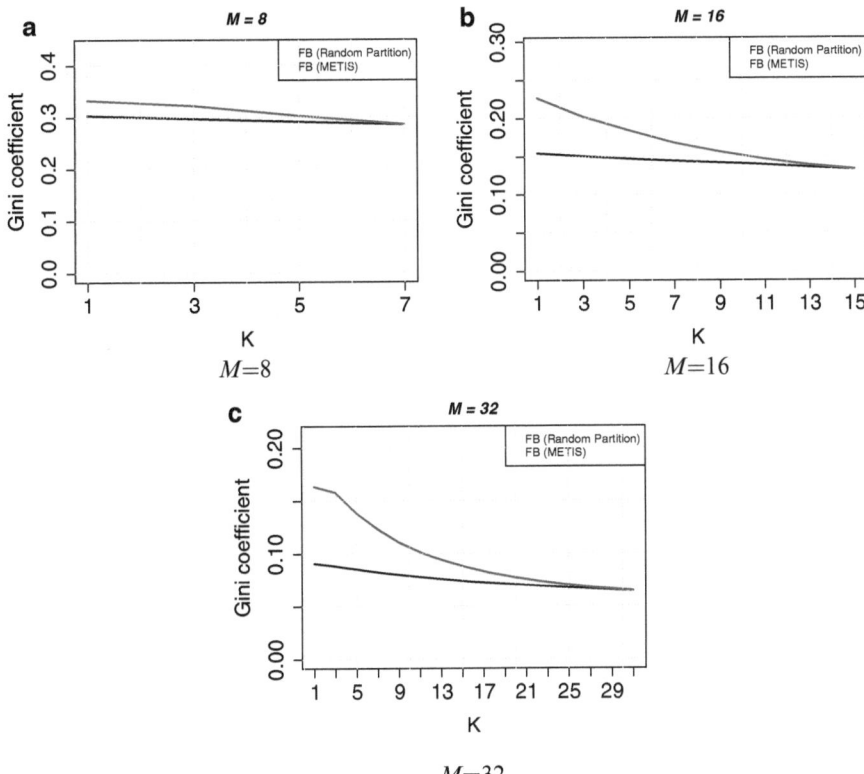

Fig. 4.4 Write load balancing in terms of Gini coefficient. (**a**) $M = 8$ (**b**) $M = 16$ (**c**) $M = 32$

while replicating data, attempts to put those socially connected into the same server as much as possible. Compared to random replication which is a de facto approach for today's most OSNs, preliminary results have shown S-CLONE to be more efficient by a substantial margin. To date, S-CLONE is the only socially aware replication scheme applicable to OSNs that offers equal data availability for every user and considers weighted links in the social graph and heterogeneous query rates in the user activity. S-CLONE can also be modified easily to work with systems that require low read load and balanced storage load instead of balanced write load.

Chapter 5
Epilogue

Social locality is a property that should be preserved in the data storage of any OSNs. In this brief book, we have presented several problem settings whose purpose is to optimize the server performance under various objectives and constraints. We have specifically focused on data partitioning and replication, two of the most important problems in storage design. Understanding that there are other issues, including data modeling, versioning, and network management, we hope that the concepts and techniques discussed in this book will serve as a building block for future development of truly socially aware storage systems.

We have presented two socially aware techniques, S-PUT and S-CLONE, as solution frameworks to the selected partitioning and replication problems. There is room for further research though. In the current development of S-PUT and S-CLONE, these techniques are invoked periodically after each update and analysis of the actual workload to accommodate changes in the user write/read rates and user-to-user social strengths. The next step will be to investigate how these techniques can adapt incrementally to changes in the social graph and user activities, so we can avoid frequent re-partitioning and re-replication procedures which can be expensive. It will also be interesting to further investigate the use of the evolutionary approach, especially for the replication problem, as this approach is naturally a good fit for multi-objective optimization. For practical use, a solution adopting this approach should be reasonably fast and adaptive to both data dynamics and network dynamics.

Replication is just one way to improve data availability. One could also use erasure codes to introduce a comparable degree of data redundancy at a lower storage cost. In an erasure code, blocks of data are combined in various ways to produce a number of encoded blocks which, instead of the original data, are distributed in the network. An important research problem is therefore to investigate whether or not preservation of social locality among the data blocks will be helpful in improving the performance of an erasure-coded storage system.

The optimization problems discussed in this book assume that the cost to send a query from a server to another is identical regardless of the communication cost between them. In practice, this communication cost varies depending on factors

D.A. Tran, *Data Storage for Social Networks: A Socially Aware Approach*, SpringerBriefs in Optimization, DOI 10.1007/978-1-4614-4636-1_5, © Duc A. Tran 2012

such as geographic locations of the servers and how these servers are interconnected in the server center (e.g., using a DHT). It is thus an interesting problem to achieve the optimization objectives reckoning with the heterogeneity of cross-server communication costs.

Social locality is not the only property that should be taken into account in data storage systems. Locality of other types such as content-based locality and geographic locality are also important. For example, the importance of geographic locality is evident in the success of Akamai to improve data access by deploying content servers closer to the users. One could argue against preservation of social locality because it may weaken other locality properties such as geographic locality. However, a nice property of preservation of social locality is that doing so will, to a large extent, also result in preservation of geographic locality. Indeed, it has been empirically shown that in an online social network geographic proximity increases the probability of friendship. Specifically, the number of neighbors of a given user decreases quickly with geographic distance and, consequently, most neighbors should stay in the local region of this user. For example, in LiveJournal.com, letting $P(\delta)$ denote the proportion of links with geographic distance δ, it has been observed in [24] that the relationship between friendship and geographic proximity can be modeled as $P(\delta) = 1/\delta^{1.2} + 5 \times 10^{-6}$. Also, it is reported in [12] that 58% of the links in FourSquare.com, 36% in BrightKite.com, and 32% in LiveJournal.com, three popular OSNs, are less than 100 km; these OSNs have average link distances of 1,296 km, 2,041 km, and 2,727 km, respectively.

To conclude, social locality should be considered one of the most important design factors in distributed social data storage and more research in the socially aware approach is definitely worth pursuing.

References

1. Adya, A., Bolosky, W.J., Castro, M., Cermak, G., Chaiken, R., Douceur, J.R., Howell, J., Lorch, J.R., Theimer, M., Wattenhofer, R.P.: Farsite: federated, available, and reliable storage for an incompletely trusted environment. In: Proceedings of the 5th Symposium on Operating Systems Design and Implementation, OSDI '02, pp. 1–14. ACM, New York, NY, USA (2002)
2. AirwideSolutions: Mobile social networking and the rise of the smart machines— 2015ad. (2010) http://www.airwidesolutions.com/whitepapers/MobileSocialNetworking.pdf. White paper
3. Arora, S., Rao, S., Vazirani, U.: Expander flows, geometric embeddings and graph partitioning. In: Proceedings of the Thirty-sixth Annual ACM Symposium on Theory of Computing, STOC '04, pp. 222–231. ACM, New York, NY, USA (2004)
4. Burrows, M.: The chubby lock service for loosely-coupled distributed systems. In: Proceedings of the 7th Symposium on Operating Systems Design and Implementation, OSDI '06, pp. 335–350. USENIX Association, Seattle, Washington, USA (2006). http://dl.acm.org/citation.cfm?id=1298455.1298487
5. Carrasco, B., Lu, Y., da Trindade, J.M.F.: Partitioning social networks for time-dependent queries. In: Proceedings of the 4th Workshop on Social Network Systems, SNS '11, pp. 2:1–2:6. ACM, New York, NY, USA (2011). 10.1145/1989656.1989658. http://doi.acm.org/10.1145/1989656.1989658
6. Cattell, R.: Scalable sql and nosql data stores. SIGMOD Rec. **39**(4), 12–27 (2011). 10.1145/1978915.1978919. http://doi.acm.org/10.1145/1978915.1978919
7. Chang, F., Dean, J., Ghemawat, S., Hsieh, W., Wallach, D., Burrows, M., Chandra, T., Fikes, A., Gruber, R.: Bigtable: A distributed storage system for structured data. In: Proceedings of the 7th Conference on USENIX Symposium on Operating Systems Design and Implementation, OSDI '06, **7**, 15–15. USENIX Association, Seattle, WA, USA (2006). http://dl.acm.org/citation.cfm?id=1267308.1267323
8. Chang, F., Dean, J., Ghemawat, S., Hsieh, W.C., Wallach, D.A., Burrows, M., Chandra, T., Fikes, A., Gruber, R.E.: Bigtable: a distributed storage system for structured data. In: Proceedings of the 7th USENIX Symposium on Operating Systems Design and Implementation, vol. 7, pp. 15–15. Berkeley, CA, USA (2006)
9. Curino, C., Zhang, Y., Jones, E.P.C., Madden, S.: Schism: a workload-driven approach to database replication and partitioning. PVLDB **3**(1), 48–57 (2010)
10. Deb, K., Agrawal, S., Pratap, A., Meyarivan, T.: A fast and elitist multiobjective genetic algorithm: Nsga-ii. IEEE Trans. Evol. Comput. **6**(2), 182–197 (2002)
11. DeCandia, G., Hastorun, D., Jampani, M., Kakulapati, G., Lakshman, A., Pilchin, A., Sivasubramanian, S., Vosshall, P., Vogels, W.: Dynamo: amazon's highly available key-value store. SIGOPS Oper. Syst. Rev. **41**, 205–220 (2007)

12. Scellato, S., Mascolo, C., Musolesi, M., Latora, V.: Distance matters: geo-social metrics for online social networks: salvatore scellato and cecilia mascolo and mirco musolesi and vito latora. In: Proceedings of the 3rd Workshop on Online Social Networks (WOSN '10). pp. 8–8. USENIX Association, Boston, MA, USA (2010). http://dl.acm.org/citation.cfm?id=1863190. 1863198

13. Ehrgott, M., Gandibleux, X.: Multiple criteria optimization. State of the Art Annotated Bibliographic Surveys. Kluwer Academic, Dordrecht (2002)

14. Ghemawat, S., Gobioff, H., Leung, S.T.: The google file system. SIGOPS Oper. Syst. Rev. **37** (2003)

15. Haghani, P., Michel, S., Cudré-Mauroux, P., Aberer, K.: LSH at large—distributed KNN search in high dimensions. 11th International Workshop on the Web and Databases (WebDB). Canada (2008). http://webdb2008.como.polimi.it/images/stories/WebDB2008/paper14.pdf

16. Hewitt, E.: Cassandra: The Definitive Guide, 1st edn. O'Reilly Media, pages 332 Nov 29, (2010)

17. Karger, D., Lehman, E., Leighton, T., Panigrahy, R., Levine, M., Lewin, D.: Consistent hashing and random trees: distributed caching protocols for relieving hot spots on the world wide web. In: Proceedings of the Twenty-ninth Annual ACM Symposium on Theory of Computing, STOC '97, pp. 654–663. ACM, New York, NY, USA (1997)

18. Karypis, G., Kumar, V.: A fast and high quality multilevel scheme for partitioning irregular graphs. SIAM J. Sci. Comput. **20**, 359–392 (1998)

19. Karypis, G., Kumar, V.: A fast and high quality multilevel scheme for partitioning irregular graphs. SIAM J. Sci. Comput. **20**, 359–392 (1998)

20. Kubiatowicz, J., Bindel, D., Chen, Y., Czerwinski, S., Eaton, P., Geels, D., Gummadi, R., Rhea, S., Weatherspoon, H., Weimer, W., Wells, C., Zhao, B.: Oceanstore: an architecture for global-scale persistent storage. SIGPLAN Not. **35**, 190–201 (2000)

21. Lakshman, A., Malik, P.: Cassandra: a decentralized structured storage system. SIGOPS Oper. Syst. Rev. **44**, 35–40 (2010)

22. Lamport, L.: Time, clocks, and the ordering of events in a distributed system. ACM Commun. **21**, 558–565 (1978)

23. Leskovec, J., Lang, K.J., Mahoney, M.: Empirical comparison of algorithms for network community detection. In: Proceedings of the 19th International Conference on World Wide Web, WWW '10, pp. 631–640. ACM, New York, NY, USA (2010). http://doi.acm.org/10.1145/1772690.1772755. http://doi.acm.org/10.1145/1772690.1772755

24. Liben-Nowell, D., Novak, J., Kumar, R., Raghavan, P., Tomkins, A.: Geographic routing in social networks. Proc. Natl. Acad. Sci. USA **102**(33), 11623–11628 (2005)

25. Merkle, R.C.: A digital signature based on a conventional encryption function. In: A Conference on the Theory and Applications of Cryptographic Techniques on Advances in Cryptology, CRYPTO '87, pp. 369–378, Springer-Verlag, UK (1988). http://dl.acm.org/citation.cfm?id=646752.704751

26. Newman, M.E.: Modularity and community structure in networks. Proc. Natl. Acad. Sci. USA **103**(23), 8577–8582 (2006)

27. Nguyen, K., Pham, C., Tran, D.A., Zhang, F.: Preserving social locality in data replication for social networks. In: IEEE ICDCS 2011 Workshop on Simplifying Complex Networks for Practitioners (SIMPLEX 2011). Minneapolis, MN (2011)

28. Niesen: Social networks/blogs now account for one in every four and a half minutes online. (2010) http://blog.nielsen.com/nielsenwire/global/social-media-accounts-for-22-percent-of-time-online/ . Report

29. Nishida, H., Nguyen, T.: Optimal client-server assignment for internet distributed systems. In: 20th International Conference on Computer Communications and Networks (ICCCN 2011). Maui, Hawaii, USA (2011)

30. Pitoura, T., Ntarmos, N., Triantafillou, P.: Replication, load balancing and efficient range query processing in dhts. In: Proceedings of the 10th international conference on Advances in Database Technology, EDBT '06, pp. 131–148, Munich, Germany, Springer-Verlag, Heidelberg (2006). http://dx.doi.org/10.1007/11687238_11

31. Pitoura, T., Triantafillou, P.: Load distribution fairness in p2p data management systems. Data Engineering, International Conference on **0**, 396–405 (2007). http://doi.ieeecomputersociety. org/10.1109/ICDE.2007.367885

32. Pujol, J.M., Erramilli, V., Siganos, G., Yang, X., Laoutaris, N., Chhabra, P., Rodriguez, P.: The little engine(s) that could: scaling online social networks. In: Proceedings of the ACM SIGCOMM 2010 Conference, pp. 375–386. ACM, New York, NY, USA (2010)

33. Reiher, P.L., Heidemann, J.S., Ratner, D., Skinner, G., Popek, G.J.: Resolving file conflicts in the ficus file system. In: Proceedings of the USENIX Summer 1994 Technical Conference on USENIX Summer 1994 Technical Conference, USTC '94, **1**, 183–195, Boston, Massachusetts, USENIX Association, USA (1994). http://dl.acm.org/citation.cfm?id=1267257.1267269

34. Renesse, R., Dumitriu, D., Gough, V., Thomas, C.: Efficient reconciliation and flow control for anti-entropy protocols. In: Proceedings of the 2nd Workshop on Large-Scale Distributed Systems and Middleware, LADIS '08, pp. 6:1–6:7. ACM, Yorktown Heights, New York, USA (2008). http://doi.acm.org/10.1145/1529974.1529983

35. Rowstron, A., Druschel, P.: Storage management and caching in past, a large-scale, persistent peer-to-peer storage utility. SIGOPS Oper. Syst. Rev. **35**, 188–201 (2001)

36. Satyanarayanan, M., Kistler, J.J., Kumar, P., Okasaki, M.E., Siegel, E.H., Steere, D.C.: Coda: A highly available file system for a distributed workstation environment. IEEE Trans. Comput. **39**, 447–459 (1990)

37. Terry, D.B., Theimer, M.M., Petersen, K., Demers, A.J., Spreitzer, M.J., Hauser, C.H.: Managing update conflicts in bayou, a weakly connected replicated storage system. SIGOPS Oper. Syst. Rev. **29**, 172–182 (1995)

38. Tran, D.A., Nguyen, K., Pham, C.: S-CLONE: Socially-aware data replication for social networks. Computer Networks **56**(7), 2001–2013. Elsevier North-Holland, Inc., USA (2012). http://dx.doi.org/10.1016/j.comnet.2012.02.010

39. Viswanath, B., Mislove, A., Cha, M., Gummadi, K.P.: On the evolution of user interaction in facebook. In: Proceedings of the 2nd ACM Workshop on Online Social Networks, WOSN '09, pp. 37–42. ACM, Barcelona, Spain, USA (2009). http://doi.acm.org/10.1145/1592665. 1592675

40. Zitzler, E., Laumanns, M., Thiele, L.: SPEA2: Improving the strength pareto evolutionary algorithm for multiobjective optimization. In: Giannakoglou, K., et al. (eds.) Evolutionary Methods for Design, Optimisation and Control with Application to Industrial Problems (EUROGEN 2001), pp. 95–100. International Center for Numerical Methods in Engineering (CIMNE) (2002)

41. Zitzler, E., Thiele, L.: Multiobjective optimization using evolutionary algorithms—a comparative case study. In: Proceedings of the 5th International Conference on Parallel Problem Solving from Nature, PPSN V, pp. 292–304. Springer-Verlag, London, UK (1998). http://dl.acm.org/citation.cfm?id=645824.668610